T0380883

When are We going to Blow Up the Volcano?

Michele Jeanmarie
Illustrated by Moch Shobaru

Archway Publishing books may be ordered through booksellers or by contacting:

Archway Publishing
1663 Liberty Drive
Bloomington, IN 47403
www.archwaypublishing.com
844-669-3957

Interior Image Credit: Moch Shobaru

ISBN: 978-1-6657-6201-4 (sc)
978-1-6657-6202-1 (e)

Library of Congress Control Number: 2024912524

Print information available on the last page.

Archway Publishing rev. date: 06/21/2024

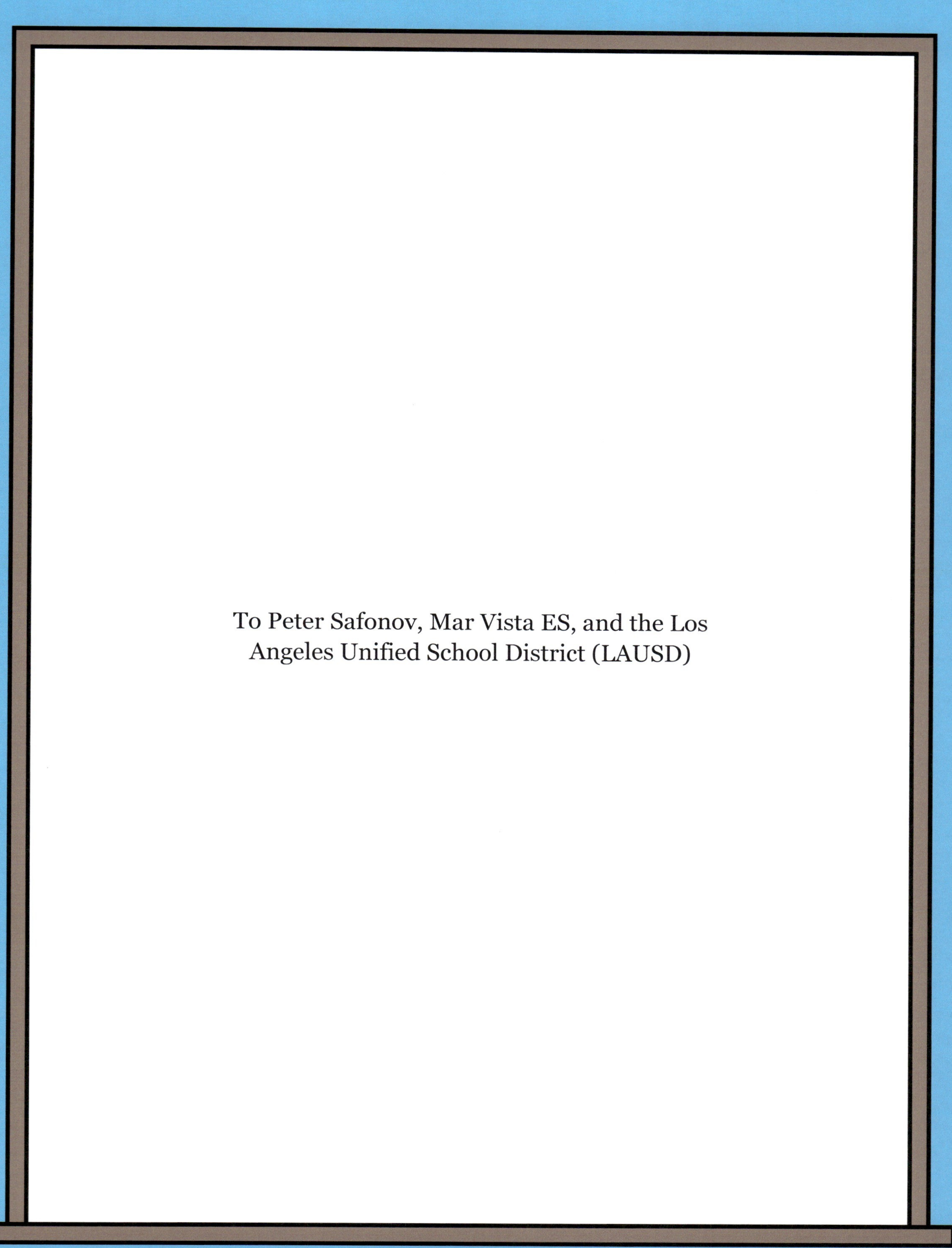

To Peter Safonov, Mar Vista ES, and the Los Angeles Unified School District (LAUSD)

We were just coming back in from lunch, when I see baking powder, baking soda and vinegar on the front table. With my head still attuned to the front, I hear classical music, and I proceed to my table, where I take out my book for Silent Sustained Reading, SSR, for short. As everyone is filing in, they too, have their eyes pierced at the front table wondering if the materials are for blowing up the volcano.

Peter was the first to inquire, “Are we going to go blow up the volcano today, Ms. Jimbaz?”

BAKING POWDER
BAKING SODA
VINEGAR

Great! She thinks. Someone asks, but Ms. Jimbaz simply smiles. We all glare at her. We proceed to our desks to read to music from Hiroshima, some jazz to calm us down right after recess.

Then the music stops. Peter asks, "Is SSR over?" For although he wants so much to blow up the volcano, it is easy for him to slip into deep thought, in reading, or any discipline for that matter. Engrossed in everything he does, science is his particular interest.

Ms. Jimbaz nods. We all slam our books shut and shove them into our desks. It is time for science!

At the overhead projector, Ms. Jimbaz places a world map on the glass. She slips a stack on the paper monitor's desk, for it is she to pass one to each of us. We then are instructed to take out our pencils, for we are about to go on a world tour, no passport required, just pencils.

We inhale, but in nanoseconds, we exhaled.

She reminds us of Italy, the boot; she reminds us of Madagascar, the movie; she reminds us of the United Kingdom, the bunny; she reminds us of Japan, the seahorse. That review under our belts, she proceeds to split up the map. Following, we cut along those lines. Followed by gluing the pieces as we thought they were back some two million years. Called Pangaea, this was the supercontinent, one big, enormous land piece.

Beautiful!

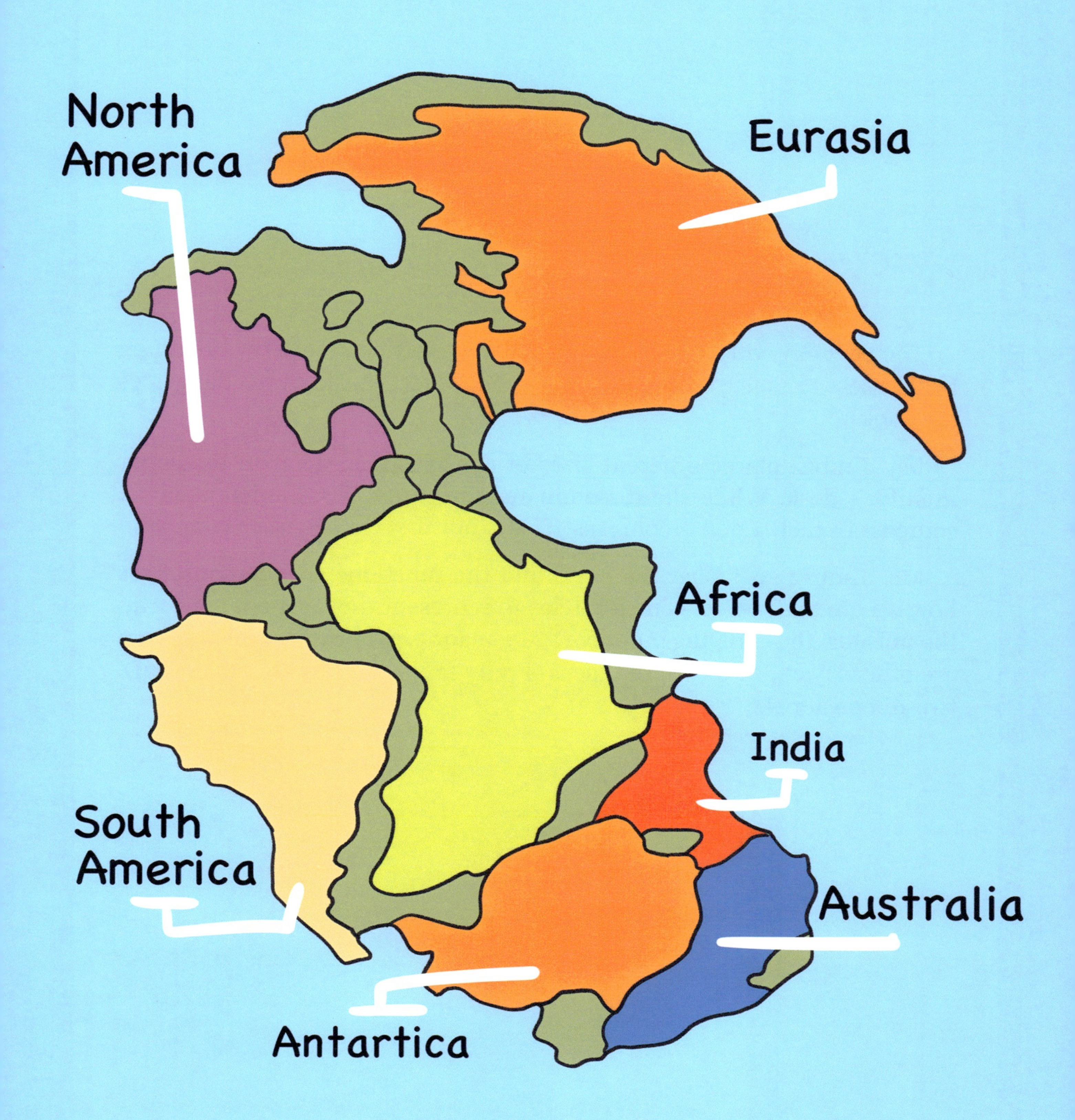
North America
Eurasia
Africa
India
South America
Australia
Antartica

The next day will be the same as the first: trotting in, heavily breathing from recess to music from Hiroshima. SSR. Click. Music stops. Bang goes the books.

Ms. Jimbaz places different sizes of construction paper on the paper monitor's desk. While she disseminates the paper, Ms. Jimbaz hands a compass to each. Cool! A compass! This is our first time to use a compass.

Ms. Jimbaz goes over the parts and the functions and demonstrates how we are to draw concentric circles to represent the layers of the earth: the nucleus, the core, the outer core, the mantle, the crust. Although there are other layers, these are the focus. Pretty fancy the way we made it 2D, two-dimensional.

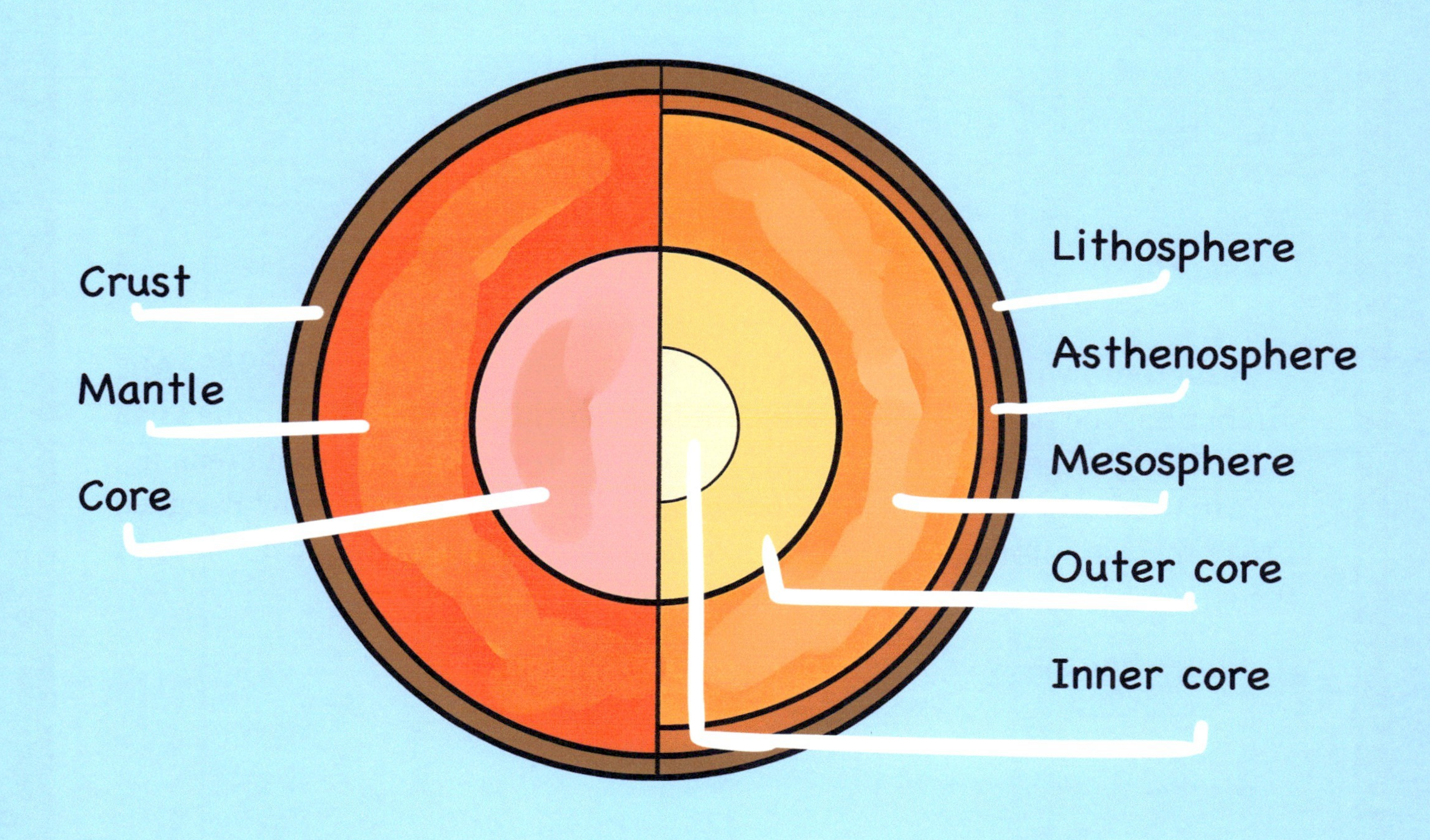
Crust
Mantle
Core
Lithosphere
Asthenosphere
Mesosphere
Outer core
Inner core

Still, there will be no blowing up of the volcano, for following SSR, Ms. Jimbaz has the paper monitor disseminate a world map. This world map has mini triangles perforated across it. These we are to crease. From creasing, we are to form a dodecahedron. Other than fragmentation to how Pangaea broke up, I am not sure what it has to do with blowing up the volcano, but if suspense is her goal, she sure succeeded.

Still, when are we going to blow up the volcano?

Another day Ms. Jimbaz starts her speech about the scientific method. She points to a three-point abbreviated chart on the wall. The three points are:

Look twice.

Work together.

Write everything down.

Ms. Jimbaz calls looking twice, “careful observation.” She calls working together, “cooperative learning.” She calls writing down, “Note-taking.”

When are we going to get started with the volcano, I wonder.

She briefly names the materials. Then she adds, “Now, if given these, what do you hypothesize? What procedures are you going to establish and follow as fellow scientists? How are you going to record your observations?” She points to a chart of the scientific method.

tific method
Look twice.
ork together.
every thing down
The Scientific Method
I. Materials:
II. Hypothesis:
III. Theory
IV. Procedure
V. Results
VI. Diagram
VII. Conclusion

No one hears a word she says.

She waits. Everyone is unresponsive. We all want to get our hands on the stuff, and she is standing in the way.

Ms. Jimbaz holds up a bottle of vinegar. This is an acid," she explains. "When it mixes with a bicarbonate" – here, she holds up a box of baking soda- "there is a chemical reaction."

Ooh, a chemical reaction!

Each table gets a tray with a box of soda, a bottle of apple cider vinegar, a bottle of distilled vinegar, a bottle of a mystery solution, a small plastic bottle with a cork in the top, and paper towels for cleanup.

Our instructions are to add two tablespoons of soda and twice as much mystery solution to the bottle, and write down what happens. We are to observe, work together and record our results. Then, we are to repeat the experiment again, keeping constant the soda, but using the other solutions. Again, we are to observe, work together and record our results.

Everyone gets closer. I hear an explosion of scream from table five. Why are they having so much fun?

"How did you guys get your corks to hit the ceiling?"

Another, "how did your bottles explode like that?"

Someone whispers, "We doubled up the amount."

I immediately quirk up. The ingredients are really undergoing some chemical reaction! Everyone is out of his seat. Oh man! I want ours to do the same thing!

Everyone looks to Ms. Jimbaz, and without a word for approval, she smiles, consenting to our childlike inquisitiveness.

Everyone is attempting one last try at the experiment. Table five is really messy. Boy! They are having a ball! Ms. Jimbaz continues, "Which combination of ingredients reacted more violently?"

Someone yells, "the distilled vinegar!"

"No! The mystery solution!" someone else contradicts when someone at the same time, posits, "What is the mystery solution?"

BAKING POWDER
BAKING SODA
5
?
VINEGAR

"Now, I warned you to take notes of your observations. It doesn't seem as if we have a consensus."

"It is the combination of apple cider vinegar and soda, Ms. Jimbaz!" someone interjects.

She repeats, "When baking soda, a bicarbonate, mixes with vinegar, an acid, a chemical reaction occurs." Then she summarizes, "This, we know by the virtue of the bubbles we see. When we mix them in an open container, we can see the bubbles, but in the bottle, sealed with a cork, in a constrained environment, the bubbles literally need some place to go, and so it explodes, causing the cork to pop up."

We are all bobbing our heads.

"So, why the confusion?" Ms. Jimbaz knowingly asks after a sigh. "You got carried away and were not taking scientific data."

The following day, as were entering the classroom for SSR, Peter asks once again, “Are we going to blow up the volcano today?”

“If time allows,” Ms. Jimbaz says, as she sheds some hope.

Everyone looks upon the table where the materials are typically set prior to science. But there isn’t any thing there. Hmm? What are we going to do today?

Ms. Jimbaz stands up when SSR is over, and we all slam our books shut. She stoops down and pulls from a shopping bag two sodas. She shakes them.

“I have two club sodas in my hands. One is hot; the other one is cold. I am going to pass them to every one of you. When you get them, shake them a bit. In the meantime, I want you to predict; I want you to hypothesize the one you think will erupt more violently... the hot one or the cold one.”

scientific method
Look twice.
Work together.
Write every thing down.
Club Soda
Club Soda

She clarifies, "While the rest of you are waiting for the bottles, I want you to get started with the scientific write up."

As I was finishing writing my prediction, the bottles reach me, ... and, yep, one is cold and one is hot!

Every one had gotten a chance to shake the bottles. Ms. Jimbaz chooses me and another fellow scientist to open them, to test our hypotheses, as she repeats her question. She says pointing the bottle at someone will be okay.

We get up from our seats. We direct the bottles towards our fellow scientists, our classmates, open them, and.... shhhh! Everyone is screaming. Everyone is simultaneously laughing with joy! Some are even summoning that the bottles be pointed towards them!

Ms. Jimbaz raises her hand for silence. We all return to our seats when she asks, "Which one erupted more violently?"

"The hot one!" we all exclaim.

"Why do you suppose the hot one erupted more?"

No one answers. We are all busy writing down the theory, as we know this was next. I am first to finish, and in an attempt to get to the volcanoes, I elaborate that, "in warmer things, such as solutions, the molecules are denser, are packed, are moving about rapidly."

Then I add, "That's why when you stir in sugar in hot tea, it dissolves quicker. Do the same in a cup of iced tea, and the sugar takes longer to dissolve. Heat excites molecules. Cold doesn't."

Ms. Jimbaz, from her schoolbag, then retrieves some more club soda, Kool-Aid and sugar. She makes us some Kool-Aid, and boy! Was it refreshing!

"And, so, when you opened the bottles, the molecules in the warmer solution, being excited in a constrained environment, burst out in an uproar!" summarizes methodically Ms. Jimbaz.

We all agree. This is so fresh! I look up and disappointedly realize that time has run out and so, there isn't going to be time for the volcanoes.

Club
Soda
Club
Soda

It is another day! As we are entering the hallway, we smell something, something familiar, but we could not tell. We hurriedly enter the classroom and see Ms. Jimbaz stirring something over the hot plate.

“What are you making, Ms. Jimbaz?” asks Peter. Are we going to blow up the volcano today?

“Fudge,” she replies. “And, no, we are not going to have time.”

“Mmmm,” some of us are savoring the air.

We anxiously anticipate more science. Ms. Jimbaz asks us to clear our tables and summons for our table captains to get a tray, a bowl, a cup of ice water and some fudge.

She starts, “Can anyone tell me how the islands of Hawaii were formed?”

No one answers.

She then continues, “Today we are going to do an experiment that helps us understand how and why we have so many islands in the Pacific Ocean. In your bowls, I want you to pour some ice water. That is going to represent the Pacific Ocean. Then I want you to add some fudge, a little at a time, to the Pacific Ocean. That is the magma, sitting on the bottom, waiting and waiting until all the parts of the volcano are formed. This is the liquid we see rising up and through a volcano.” She then asks, “What is magma called once it’s outside the volcano?”

“Lava!” some shout.

“Then I want you to carefully observe what happens to the lava, to the fudge,” she continues.

We begin. We draw in closer. We watch as the lava hardens in the cold water. Then we share our results. We quickly write down our observations.

Ms. Jimbaz summarizes, “The big island of Hawaii has an active volcano. When it erupts, it spews out magma, which becomes lava. Lava, in ocean waters, hardens. Over a period of time, lava piles up on top of old lava until it reaches and surpasses the water level and becomes an island.”

A student, who travels to Hawaii, confirms, “That’s why there are so many islands over there. We took a tour.”

Sugar
CHOCO

Everything begins to make sense. We finish our scientific write up. To our icy bowls of fudge, we add milk. The result: chocolate fudge milk. How tasty!

Hmm.

Nothing is on the table, but as soon as SSR is over, Ms. Jimbaz has the paper monitor pass out drawing paper. We are instructed to sketch a volcano to include all parts: the cone, the magma chamber, the lava flow, the ash cloud, the crater, the lava flow. Although we know the importance of labeling, we are reminded.

As we are drawing and coloring, Ms. Jimbaz sets out an industrial sized bag, called Plaster of Paris. She scoops out a tub full, places it on the trays, adds paper towel tubes, cardboard and tubs of water, red and orange paint, and paint brushes.

Of these materials, we guess the function and purpose of each. We set out the sequence, the order to build a three-dimensional volcano. We get to work. We clean up. We leave our volcanoes to dry overnight.

The end of the week is here! The baking soda, the baking powder and the vinegars are sitting once again on the front table. Cool! We are about to undertake another exciting experiment!

The volcano? The volcano? Could it be the volcano? It had been a week since we have made the volcano! I look around. Oops!

"Ms. Jimbaz, are...we... go...ing... to... b-low... up... the... vol-ca-no... to-day?" I heard later Chela syllabicated.

Smilingly, she corrects the terminology, "Today ... we're going to ... not... blow up the volcano.... but .... make it erupt."

Peeeeeeeeeter!

“Where’s Peter?”

“Where’s Peter?”

“Where’s Peter?”

“Oh, no!”

Peeeeeeeeeeeeeeeeeeeeeter!

scientific method
VINEGAR